Praise for Infinite Gamification

"GREAT. Concise, easy to read, complete... Great work."

Nicolas Babin

"In one word, it's EXCELLENT. I really found it to be informative and a useful guide to help introduce sustainable performance management techniques into an organisation."

John Conaghan

"Toby Beresford's Infinite Gamification is one of the best references I've seen on score-driven motivation systems."

Yu-Kai Chou

Infinite Gamification

Motivate your team until the end of time

By Toby Beresford

First published 2020 by Bot Kiln

ISBN 978-1-8380453-3-3

Edited and typeset by Bot Kiln www.botkiln.com

CONTENTS

FOREWORD

Most books or courses on gamification tend to focus on the broad picture, either introducing surface level concepts or presenting as many approaches as possible to the gamification process.

In Infinite Gamification, Toby Beresford focusses on a specific experience in gamification and then offers practical techniques for maximising the experience.

Most of us in the gamification community tend to focus on "finite gamification," programs that have a beginning, middle, and end. But Beresford notes that "infinite gamification" is powerful in situations where consistent improvement over time is desirable, such as in the workplace. Instead of completing a training course, for example, and then getting back to work, infinite gamification efforts seek to improve performance in lasting ways by motivating players day after day for weeks, months, perhaps even years.

The main mechanism Beresford examines is the use of scoring. He shares with readers why and how scoring has sustained motivation. He then offers different metrics we can use to engage players, as well as different methods for scoring. Next he offers practical advice about adverse situations we may face in infinite gamification. Finally, and perhaps most enlightening for me, Beresford presents the "Maturity Model" whereby players have different attitudes toward, and engagement in, gamified programs.

Even a gamification novice will find this book helpful, although the reader does need a basic understanding of gamification. However, gamification practitioners and senior leaders will find this book both applicable, and enlightening.

Jonathan Peters, PhD
CMO, Sententia, Inc.
Author, "Deliberate Fun: A Purposeful Application of Game Mechanics to Learning Experiences"

INTRODUCTION

In 2011, the word *gamification* made the Oxford English Dictionary shortlist for word of the year.

Since then, it has become one of those concepts that many business leaders feel they understand, yet too often create programs that are too complex or mix up the motivational goals of their players.

Perhaps this accounts for why so many gamification projects fail.

Finite and Infinite Gamification

Gamification can be split into two types—finite and infinite.

A **finite** program has an end, an epic win, a top level, a complete collection of badges.

An **infinite** gamification program is designed to continue forever.

There is no end, no final level to reach, no epic win, just a continuous, upward progression or an ever-greening competition.

Most gamification programs are directed by a training goal of some kind and so usually fit into the finite gamification model: there is an end when the learner passes the final hurdle to supposed mastery.

But, real life mastery doesn't happen just because you've completed the education process: there is still the need to *adopt* the skills into your day to day life and then, once adopted, *perform* them well.

It is in these latter stages—**adoption** and **performance**—that infinite gamification becomes more important.

A sustainable program that motivates teams and individuals to continuous improvement is the goal of infinite gamification.

Examples of Infinite Gamification

You may be thinking, what obscure branch of gamification is this? But, in fact, modern culture is filled with examples of infinite gamification. Indeed, some of the most successful gamification programs are infinite by design.

Good examples of infinite gamification include:

- **The Oscars**—where status is awarded annually based on the votes of a select few.
- **TIME Magazine's Person of the Year**—where status is accorded based on media coverage.
- **English Premier League** (or insert your country's favourite sporting league here)—where status is based on the results of games played throughout a season.
- **Transparency Index**—where countries compete not to be shamed at the bottom of Transparency International's leader-board.

- **UNDP Human Development Index**—where countries compete to provide an holistic standard of care for their citizens.

In fact, in most businesses you can find examples of infinite gamification such as:

- Employee of the month
- Quarterly Sales Leaderboard

Online communities also get in on the act such as:

- StackOverflow Score

And last, but not least, loyalty programs that show no sign of ending:

- Air Miles

Infinite gamification programs can be immensely powerful. They can create value for managers, teams and players. They can spawn an entire industry with an associated ecosystem around it.

When you take into account ticket sales, TV sponsorship, player salaries and so on, the English Premier League is worth around £2.5 billion a year. Not a bad outcome for a gamification design!

Who is this book for?

This book is targeted at any leader and manager seeking to influence the behaviour of those around them.

So that could mean:

- A **sales manager** looking for better sales results from your team
- A **change manager** keen to see staff use new tools and processes
- An **operations manager** with field engineers not using the new IT system properly
- A **diet and exercise coach** looking to focus participant attention on the metrics that matter
- A **charity leader** looking to influence organisations to bring in sustainable supply policies
- A **marketer** enabling customers to track success in using a particular product or service

- A **community manager** devising feedback systems that publicly recognise the most engaged community members

It doesn't matter whether your program is small scale or huge; whether it stands alone or is embedded in a larger program; whether informal or formal; or whether those you are influencing are individuals, divisions, organisations, or even entire countries.

Infinite gamification will equip you with a powerful new influence tool.

Why gamify?

Leadership, whether you are influencing colleagues or directing subordinates, usually involves an infinite gamification program of one kind or another.

At some level, leaders signpost the behaviour they want to see in others. It might be as simple as 'doing or not doing' (a binary score) or much more complicated ('doing or not doing well') with graded levels of achievement.

In return we are all conditioned to calibrate our behaviour to the score given to us. Formal education tells each of us, right from a young age, to pay attention to scores. We are all hypersensitive to our score.

This conditioning presents a problem for leaders in that even when you don't offer up an explicit score, your people will invent one of their own and channel their activity in that direction!

It is important to remember that since every organisational context is unique, with different goals and current priorities, the score that each leader highlights also will be different.

EXAMPLE:

Sales manager selling an entirely new product—key activity metric: Outbound calls made to new prospects

Sales manager selling an old product—key activity metric: Outbound calls made to previous customers and old prospects

In infinite gamification design there is no one-size-fits-all 'correct' answer. Every context is unique; each team can be at a different stage of development, and goals can vary greatly between businesses, and even teams, in the same business.

It all starts with a score

Most infinite gamification programs are based on a score.

Scores are at the heart of the most important stories we tell ourselves.

We use them to track our success at every level, whether personal or corporate.

For example:

- How many hours did I go to the gym this week?
- Is my diet working?
- Did our team improve its productivity last month?

Whether a journey takes a single step or a series of steps to reach our final goal, we usually use some sort of score to evaluate our progress:

- 47 likes for that Facebook post
- A sales target might be displayed as 57% achieved

Scores can take many forms; but they don't always look like numbers.

For instance:

- Driving over the speed limit—an unhappy face emoji.
- Volume of applause after a theatre performance, or indeed, a standing ovation!

However they are displayed though, scores can measure our own performance or that of our team, organisation or country.

Scores at best, drive personal enquiry—"did I do better this week than last week?" while at worst, they demotivate—"I will never be as good at this as them."

Whether consciously or not we all keep track of the scores that matter to us; we all operate our own scorecards.

These scorecards might look like a quick mental check "Am I doing okay at this?" or a more structured approach—a dashboard on a mobile phone, for example—"how am I doing against my targets?"

For leaders it's important to remember that if you don't show your people the score, they will probably invent one. How many of us have worked in offices where success was thought to be measured in amount of overtime worked?

With the rise of the digital world, we are now presented with more scores than ever. How many followers do I have, likes for that post, answer views this month, etc.

Scores are everywhere

These presented scores are often displayed in ways that make comparison easy. For example, analytics dashboards that compare us with previous weeks (e.g., follower increase over time), or league tables that show how we are doing versus others (ten most viewed writers).

Infinite gamification is everywhere

But these presented scores don't appear out of nowhere. Somebody, somewhere has designed the score we are using—and, of course, that means there are well-designed scores and badly designed scores.

A **well-designed** score enables you to succeed.

A **badly designed** score drives negative behaviours – like apathy, cheating, or other wrong behaviours.

Sometimes, badly designed scores can wreak havoc.

For example:

- Daredevils taking selfies from roof tops risking death in the quest for more social media "likes".

- The 5,600 Wells Fargo staff who lost their jobs in 2016 having created two million fake accounts trying to hit their sales target.

So, if a badly designed infinite gamification program can kill you, or lose you your job, it's worth taking the time to design the score well…

Book structure

Any new program comes to fruition in three main phases each outlined in this book:

- Analysis

- Design

- Evolution

I recommend you use this book as a travelling companion on your own journey. Use the models and checklists to test your own design thinking.

ANALYSIS

In the analysis phase we clarify the requirements for a new program, and we identify existing infinite gamification programs, scorecards and leagues already in place for our target players.

Prime Directive

Most infinite gamification programs start with a reason for their existence—a "prime directive".

The infinite gamification program exists to achieve the prime directive.

A clear definition of your prime directive should take no more than a sentence.

"The prime directive of this program is to...."

- Help you achieve your weight loss goals (Weight Watchers)
- Emphasise that people and their capabilities should be the ultimate criteria for assessing the development of a country, not economic growth alone. (Human Development Index)
- Increase sales through higher performance by encouraging each other with healthy competition (a sales scorecard)
- Measure social media influence (Skorr)

It is worth noting that the prime directive is often determined by the person(s) paying for the program. What gets scored, gets done.

Data Sourcing

Scores feed on a stream of raw data. Data collection is hard, and it is usually messy!

What raw data is available to you and in what format? Can the data collection be automated?

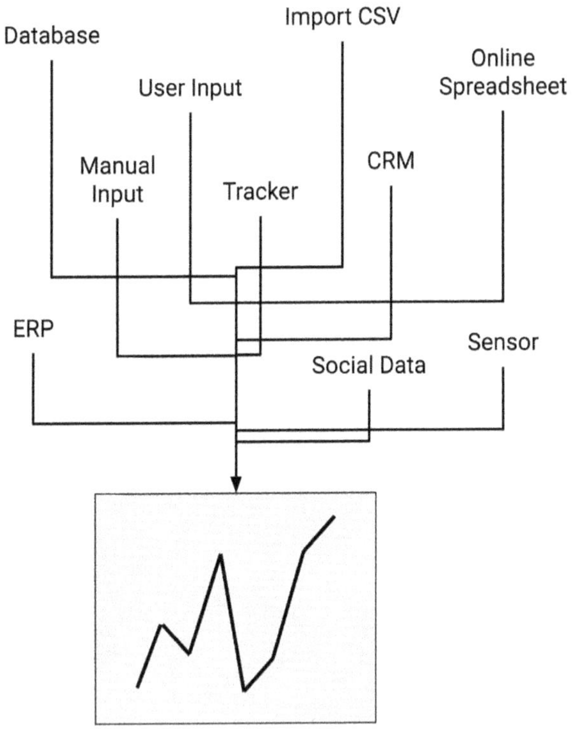

EXAMPLE:

Field engineer score at an electricity grid management company

Prime directive: Improve usage of new tablets for entering job data

Data sources available:

ERP (Enterprise Resource Planning) systems allow for data to be exported via views. These views have to be set up for the score metrics needed.

Score metrics required:

- Travel time to job site from job assignment
- Time from job start to job completion
- Number of jobs completed successfully
- Number of jobs postponed or passed to another team

Data journey:

In this example the data is passed from site to the ERP (via the tablet interface) and then redisplayed as a personal analytics score back to the field engineers.

Benefits:

The field engineer can see when data has been entered incorrectly (e.g., jobs that only took an hour yet appear on the system as unfinished).

Management goals (e.g., reduced elapsed time between job assignment and arrival at the job site—aka 'putting down your cup of tea and leaving immediately') are made more visible to field engineers.

The infinite gamification design means that no further external benefits or remuneration are required—the program can continue indefinitely.

Score Context

Since scores and infinite gamification programs are already everywhere there is no such thing as a "green field" site. You are not introducing your score into a vacuum.

Understanding score context means asking how your planned program is positioned with regard to the alternative scores your audience already has available?

For example, say we want to give software developers in our team a score. First, we need to review the scores they are already watching. For example, they might have broad, informal scores such as Reddit Karma points, or more narrow, specialised scores such as StackOverflow points.

Formal scores are the next level of importance because they are usually attached to wider benefits or responsibilities. They include salary and seniority—broad measures shared with other staff.

Narrow formal scores are the most critical (if they exist) because they are a formal measure of our own specific performance. For an internal developer this might be percent of target sprint tasks achieved; for contract developers this might be the number of hours they work each week.

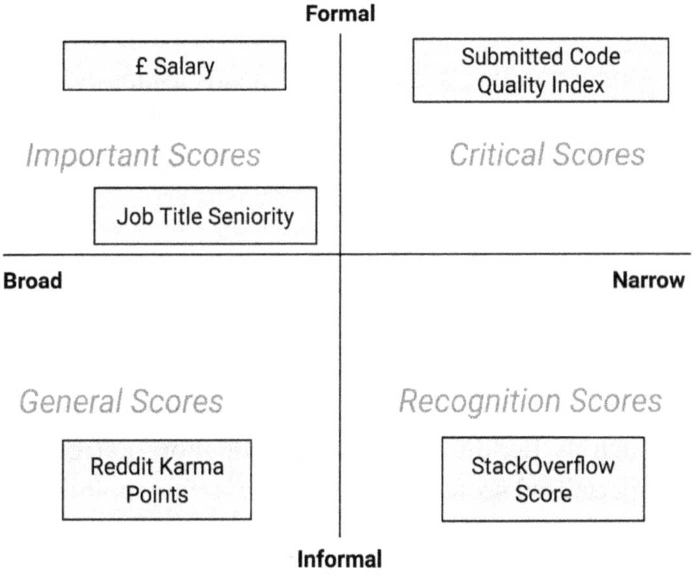

Stakeholder Analysis

What does each part of the surrounding community expect from the score?

An onion diagram is a great way to map the community that is affected by the program and to think through its unique needs.

In this example we've mapped the stakeholders for a simple sales leaderboard:

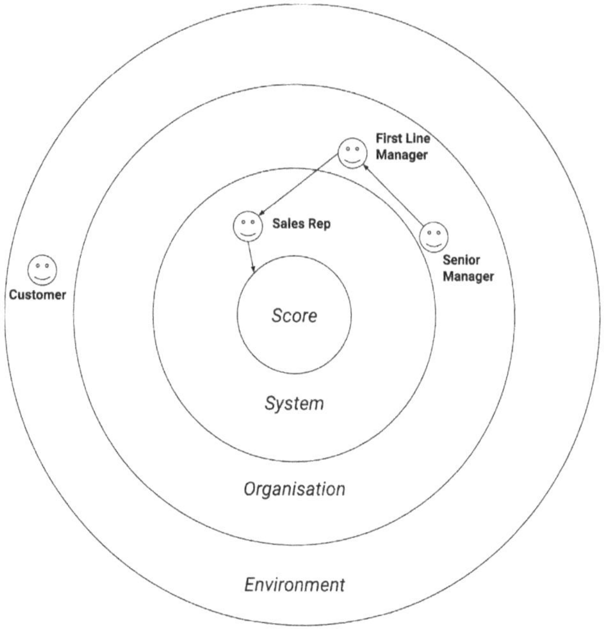

Here, at the heart of the infinite gamification program is the score which is rendered as a personal score to the sales rep—percent of target achieved this week, for example.

The first line manager will be more interested at an organisation level—seeing the percent of target achieved by each rep in their team on a leaderboard.

A senior manager might want to simply see the aggregate of target achieved, so a single number is all they need.

Finally, outside the organisation is the customer who is affected by this leaderboard—they certainly don't want to be called twice a day by reps trying to hit their targets.

Infinite Gamifier Types

When designing programs, the role you play when you publish a scorecard for your players is worth bearing in mind.

There are five potential roles to play:

- **Manager** - looking for organisational outcomes
- **Coach** - looking for user outcomes
- **Commentator** - looking for audience engagement
- **Referee** - maintaining a level playing field
- **Player-Coach** - extending a self-tracking service to others

As an infinite gamifier you will usually pick one of these roles and stick to it throughout the program.

EXAMPLES:

Manager - an operations manager reporting a field engineer score for use of new software system at an electricity grid company or, a sales manager looking for sales revenue increase through higher call volumes.

Coach - an internal sales trainer upskilling staff by tracking digital selling inputs and adoption of key tools and activities (e.g. rep page visits, social media posts, and comments).

Commentator - blogger publishing university league tables based on student satisfaction, graduate job prospects, and research results (e.g. number of articles published in journals per year).

Referee - impartial trainer running a cross-fit box league competition with an overall winner.

Player-Coach - participant and organiser of a weight loss program on a WhatsApp group, tracking number of pounds lost each week, with a weekly weigh in for all participants.

Player Needs Analysis

Finding out what your player needs from the program is very important—after all, they will be the ones seeing the score and league day to day. Your players will be expected to change their behaviour as a result of the score. If it doesn't work for them, then behaviour will not change.

Player needs analysis is an ongoing activity—you need to keep checking in to see what is, and isn't working.

Tools at your disposal include:

- Focus Groups
- One-on-one interviews (phone, face to face)
- Net Promoter Score Questionnaires
- Historical Data Analysis

EXAMPLES:

- **One-on-one interviews** - these are the most useful. Honest conversations really help you see how the program is being received. If you're running a pilot, it may bring out negative feedback that can help with framing[1] the main program correctly. Comments such as 'is this just another way for management to keep track of us?' will need careful consideration.

- **Focus Groups** - getting a group of players together helps them find areas of common concern. For example, 'our social media influence score should be based on data from multiple social media channels—not just blogs and twitter. We want it to include Instagram and Facebook too.'

- **Net Promoter Score (NPS) Questionnaires** - a quantitative score can be a helpful tool to track and measure your own progress. Baseline when you circulate your program design, immediately after launch, and then a regular poll of NPS score will help you see whether your program is maintaining its popularity.

- **Historical Data Analysis** - by tracking the number of phone calls and

[1] See "Program Framing" later in this book for more.

sales booked that your reps make each week, you might discover patterns such as a week with poor sales often preceded by a week with a low volume of outbound calls.

If you are working with a broad audience, for example, cross-functionally, you may need to segment your players into different **player personas**. Player needs can then be segmented accordingly—for example, by how quickly they adopt new ideas, or by how motivated they are by factors such as achievement, competition, or socialising.

"If you can't measure it, you can't manage it."

Player Personas

Analysing the detailed personas of your players improves program adoption by ensuring you view the program from their point of view.

There are many different ways to model user types and create personas, this is my own list of questions to consider.

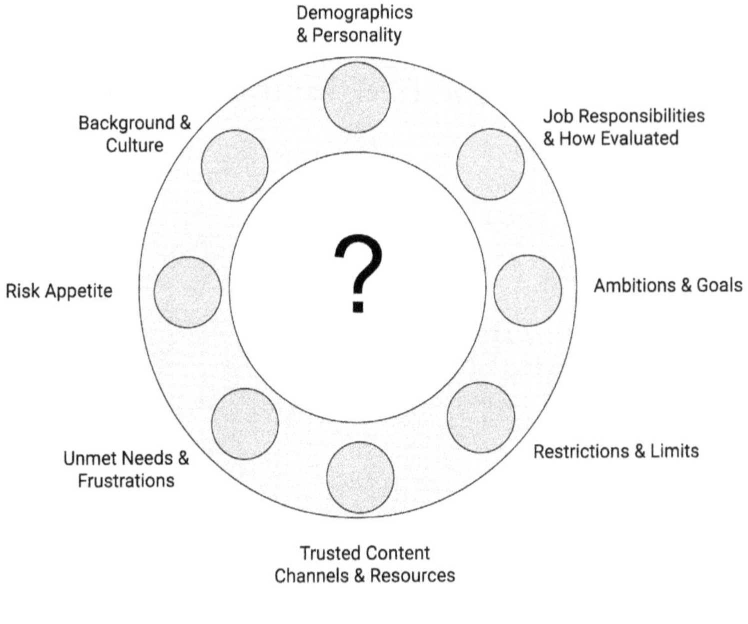

If you do find you have different personas, your design will inevitably be more complex because your program will not motivate everyone in the same way.

However, here's an example of a player persona developed for an infinite gamification program where there was one job role—the sales rep:

?

Our player persona's name is "Achiever Andrew".

Job Responsibility and How Evaluated

Andrew is an inside sales rep responsible for booking new business from existing clients and handling inbound sales requests from new clients. Evaluated on percent achieved sales revenue against target per quarter.

Ambitions and Goals

Key ambition is to over-perform against quota target and unlock bonus commissions on top of usual salary, reflected in a desire for a new sports car.

Restrictions and Limits

Andrew is responsible for a particular product line and for clients calling within a specific geographic area.

Trusted Content Channels and Resources

The corporate intranet "seller bible," the external facing website with product descriptions, and monthly team meetings are the most important. Andrew also listens to sales training podcasts on his way to work and reads a free newspaper on the train each day.

Unmet Needs and Frustrations

Andrew does not have sufficiently senior relationships with some key clients to grow the client account. There is a lack of visibility of client business plans.

Risk Appetite

Andrew tends towards a low risk approach, preferring tried and tested sales techniques to trying something new.

Background and Culture

As an inside sales rep, traditionally the role has been about waiting for client service orders rather than reaching out to proactively find clients and new sales.

Demographics and Personality

Andrew is a typical rep. Our reps are mostly aged 24 – 35 with a 'type A' achiever personality. Winning against peers is often perceived as more important than overall team success.

"Human beings have an
innate inner drive to be
autonomous,
self-determined, and
connected to one
another.

And when that drive
is liberated, people achieve
more and live richer lives."
— Daniel H. Pink

Player Motivation

Most of us are conditioned to believe that it's worth working toward a higher score. But, as infinite gamification score designers we need to communicate and manage the rewards associated, in order of cost, with the cheapest to deliver first:

Intrinsic

This is usually the best incentive of all—i.e., when the score user has a personal reason to improve their score.

Status[2]

Social Status is worth having--we get bragging rights, and can bask in the warm glow of peer adoration.

Access

Higher scores can give exclusive access to content and experiences.

---

[2] Adapted from the SAPS model - Gabe Zichermann and Christopher Cunningham. Gamification by Design: Implementing Game Mechanics in Web and Mobile Apps. Sebastopol, CA: O'Reilly, 2011.

Power

A higher score can give you power over other users.

Stuff

Most of us like shiny objects. You can incentivise with stuff—even better if it's relevant to the activity.

Cash

Cash is the ultimate extrinsic incentive, and is used most often. Be wary though, if you're giving out cash, then you're really employing people to work for you and so might have employment law and tax implications.

EXAMPLES:

Intrinsic

Someone who wants to lose weight is a willing participant in a diet and exercise program.

Status

Participants in an annual "top influencers" list, such as Forbes' '30 under 30' can enjoy a higher status than others who didn't make the cut.

Access

Winning a backstage pass or after party invite in a competition.

Power

StackOverflow users with higher scores can edit the online answers of other users.

Stuff

A health insurer offers a weekly free coffee to members who demonstrate they've been active using a smartphone app.

Cash

Salary for work performed as expected.

Player Opt-In Spectrum

Only opt-in gets buy-in. To be a player in anything is defined by choice—you have chosen to play.

The spectrum of opt-in strategies are:

- **Opt-in to Join** - player must join to opt in and may opt out at any point.
- **Granular Opt-in** - player opts in to specific features—for example to receive a personal score, but opt outs of public sharing or league comparison.
- **Opt-out** - player is opted in by default but may opt out at any point.
- **Mandatory** - player may not opt out. This is typical in workplace programs where tracking comes with the role—e.g. sales reps would expect sales performance to be tracked.

For long term success, try to avoid mandatory programs if you can: e.g. accounting firm PWC has been running global social media leaderboards that are fully Opt-In for several years.`

DESIGN

Scorecards and **leagues** are where the 'rubber hits the road' in an infinite gamification program. This is how players get feedback on their progress.

Designing a **scorecard** for each player will require you to consider the metrics—the score algorithm that combines metrics into a final score for each player.

A **league** design determines how players compare progress with others.

Vanity and Clarity metrics

Vanity metrics are the metrics other people use to evaluate you.

Clarity metrics are the operational metrics that you need to get right to be successful.

A scorecard may include vanity metrics, but usually the player can only indirectly affect them.

On the other hand, clarity metrics are the ones a player can affect directly; and so are worth an optimisation focus.

EXAMPLES:

Vanity metric - the number of Twitter followers you have.

Clarity metric - the percentage of your Twitter followers clicking on a link you shared.

"Let them talk; I have new metrics now."

— John Mark Comer

Lag and Lead metrics

Lag metrics happen after the fact—they are usually easy to measure, but difficult to influence directly.

Lead metrics are what happens beforehand—often they are the activities we can influence, but they can be hard to measure.

Lag indicators are sometimes known as "outcome" measures, while lead measures can also be called "process" measures.

If in doubt, it's often better to count the fruit not the leaves—what you actually want more of, "the fruit", rather than what gets you there, "the leaves".

EXAMPLES:

If the percentage of Twitter followers clicking on a link is the **lag** metric then the **lead** metric might be number of Twitter follower bios read per day. The reasoning being that the more you know your followers, the more likely you are to post links that interest them enough to click.

Another example, losing weight isn't an activity, but eating and exercise are. So, the **lag** indicator is weight lost; the **lead** indicators are calories consumed and exercise done.

By splitting lead and lag indicators you can easily display who's climbing—hopefully on their way to success—and who's cruising—possibly on their way to failure.

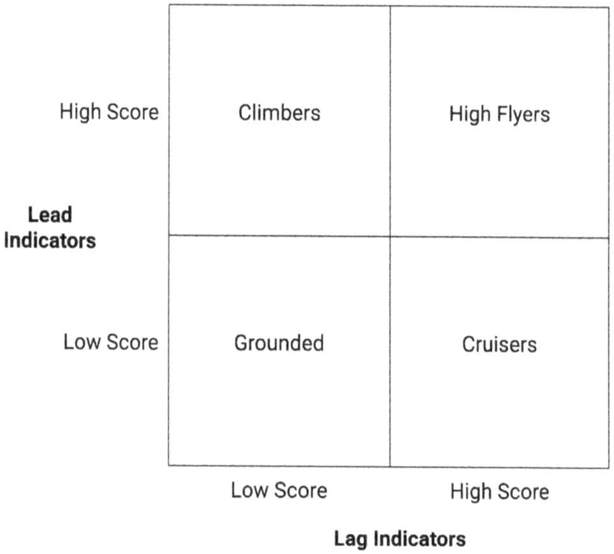

"Correlation does not necessarily imply causation."

Activity and Reciprocity metrics

Activity[3] metrics are those that the player does.

Reciprocity[4] metrics are those that occur as a direct result of the player's activity.

A good infinite gamification program typically has metrics of both types (see Metric Maturity for how the split typically changes over time).

EXAMPLES:

Activity metrics for a sales program might include number of phone calls made, volume of emails sent.

Reciprocity metrics for a sales program might be new leads signing up, a sale closed, or a positive NPS received from the customer.

---

3 Also known as 'Labour Driven'

4 Also known as 'Performance Driven'

Positive and Negative metrics

You can either count what's good or count what's bad. Generally, human beings are more motivated keeping track of the positive rather than the negative.

"There's a magic that happens when we concentrate on the good that is being done and encourage that kind of performance"[5]

— Charles A. Coonradt

EXAMPLE:

Instead of "number of injuries" many building sites use "number of injury free days" when tracking health and safety statistics.

[5] Coonradt, Charles A, Lee Benson, and Inc Game of Work. *Scorekeeping for Success*. Park City, Utah: Game of Work, Inc., 1998.

Metric Priority

Not all relevant metrics are relevant all of the time. Only analyse what you plan to optimise.

Prioritising metrics means we only need to focus on one metric type at a time. Then we progress once we have mastered that metric.

EXAMPLE:

When running a social media channel the G.E.R.M. Model provides a framework to prioritise media metrics according to the development stage of the activity:

G = **Getting going:** is our content being provided/updated regularly (e.g., number of tweets per week)?

E = **Engagement:** is our audience attentive to our content (e.g., average number of retweets per tweet)?

R = **Reach:** is our audience growing (e.g., number of new followers per week)?

M = **Monetisation:** are we monetising our audience (Conversions, Revenue earned)?

Metric Category

Categorising metrics allows you to manage them effectively.

For instance, when choosing software user metrics, you need to pick those that drive the behaviours you want:

- **Learning**: new behaviour and associated tools/ processes (e.g., emails sent).
- **Adoption**: a regular habit around the new behaviour (e.g., emails sent per day).
- **Reinforcing**: look to maintain an existing behaviour (e.g., email response rate).
- **Performing**: improved outcomes due to the target behaviour (e.g., sales meetings arranged).

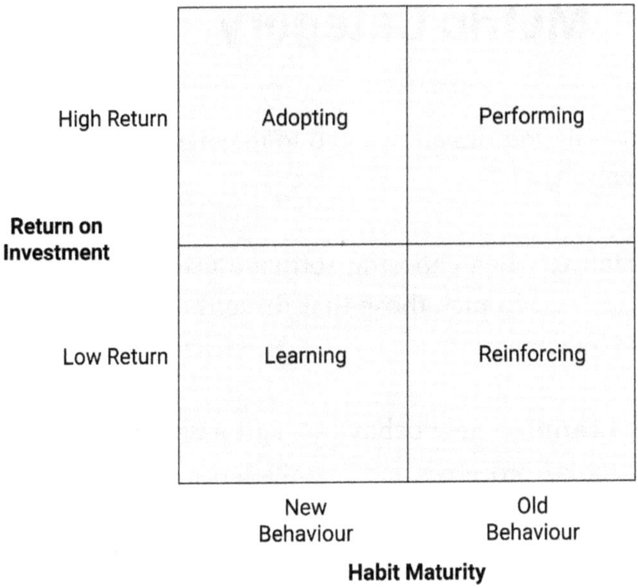

Since most players seek to optimise their score, where you allocate the majority of the score will also determine the level of return you get on the program—i.e., *performing* metrics provide the most return. This has to be balanced against making the program engaging enough for newcomers to feel they are 'getting somewhere'—hence the need for some of the score to be allocated to *learning* metrics.

Score Periods

Scores can be collected (and reset) over a time period. Score periods can be:

- All-time
- Annual
- Seasonal
- Quarterly
- Monthly
- Weekly
- Daily
- Hourly

EXAMPLES:

- **All time** - total Stack Overflow points earned.
- **Annual** - income for this tax year.
- **Seasonal** - Arsenal F.C. points in the premier league
- **Quarterly** - South West Region sales figures

- **Monthly** - Number of new logos signed this month.
- **Weekly** - Top prospect finders this week.
- **Daily** - Stock Market Index change for the day.
- **Hourly** - active minutes in the last hour.

"Not everything that counts
can be counted and,
not everything that can be
counted, counts."
—Albert Einstein

Score Methods

When you compile a score, you need to calculate it from raw score data. There are several mechanics for arriving at a score.

Here are six common ones.

1. Sum: 1,3 = 4
2. Difference: 1,3 = 2
3. Latest: 1,3 = 3
4. Max: 1,3 = 3
5. Min: 1,3 = 1
6. Average: 1,3 = 2

A score algorithm will use one of these score methods for each metric; these are then combined to make the total score.

EXAMPLES:

- **Sum** - total number of loyalty points earned
- **Difference** - new YouTube subscribers this week

- **Latest** - Current weight
- **Max** - Longest throw, from 3 attempts
- **Min** - Fastest of three attempts
- **Average** - average sales revenue per team member.

"You get what you inspect,
not what you expect."
IBM maxim

To Compare or Not to compare

As social beings we all compare, all the time. We compare our current selves with our former selves, who we think we are, and who we actually are. We compare with our neighbours, our friends, and celebrities on TV. Unless we live in a dark cave, comparison is a fact of life.

As infinite gamification designers our job is not to wipe out comparison altogether, that's just not possible, but to signpost the healthiest form of comparison for the context.

For example, a good sports coach will encourage their players to "be the best they can personally be" rather than being better than the competition.

When we set up our program, we don't have to show every possible form of comparison. We can restrict the views to the ones that matter and drive positive, encouraging, sustained engagement.

Comparison Options

There are five main ways to offer users a comparison when it comes to their scores.

- **Individual versus target**: for example, you've hit 100% of your target this week.
- **Individual versus time**: for example, showing progress against self, week to week.
- **Individual versus individual**: showing the score as it relates to the scores of peers.
- **Group versus time**: showing the score as it contributes to group performance over time, e.g., in reaching a set goal.
- **Group versus group**: showing the score as it contributes to the group performance versus other groups.

"It's not only how you perform,
but how you compare."
—Charles A. Coonradt

League Types

When comparing individuals or teams there are different ways to run a league competition.

- **Leaderboard** - players try to get the best personal score, and are then compared with others.
- **Knockout** - players play each other, with the loser being eliminated in each round until a final decides the winner.
- **Ladder** - players compete via ad-hoc challenges, with the winner taking the place of the loser.
- **Round robin / Box League** - every player competes with an individual bout against each other player in the box.

"Comparison is
the thief of joy"
—Theodore Roosevelt

Ranking Methods

Within leagues there are different ways to create a ranking.

Sum

The total of all metrics gives the total score.

Relative

Each player is ranked according to each metric and assigned a percentage of the points available for that metric, relative to their rank. For example, the first ranked player gets 100% of available points for that metric.

Rank Score

Each player is given a score according to his/her rank. For example, first ranked player gets a score of 1, second ranked player gets a score of 2, and so on. This is used in many CrossFit leagues.

Serial

The first metric takes precedence; the second metric is used only if two players have the same score in the first metric, and so on.

EXAMPLES:

- **Sum** - football leagues
- **Relative** - UNDP Human Development Index
- **Rank Score** - Crossfit leagues
- **Serial** - Olympic medals table

Common Pitfalls

Watch out for design pitfalls that reduce engagement or drive unwanted behaviours.

Engagement Reducers

- **Over Complexity**: most programs start simple
- **Irrelevant Incentives**: the reasons to participate are not strong enough
- **Score Fatigue**: too many scores and they lose their impact
- **Manipulation**: forced into activities just to get a score
- **Deceptive**: pretends to be fun when in fact it's still hard work
- **Competitive People Only**: not everyone wants to win
- **Indifferent to privacy concerns**: not everyone wants to be seen
- **First Mover Unfair Advantage**: late arrivals can never catch up

- **No rule book**: players want clarity in how to score

Unwanted Behaviours

- **Demotivated Segments**: core and low performers feel they cannot win
- **Prospering Cheats**: insufficient crack-down on cheating
- **Over-Exertion**: too much of a good thing
- **Moral Drift**: immersion means players break external rules
- **Belittling**: untracked activity now seems pointless
- **Exclusive channelling**: unscored activities are ceased altogether
- **Addiction**: playing the game too much causes you harm
- **Over Justification**: rescinded rewards now demotivate previously self-motivated activities
- **Compers**: activity just for the reward alone

Late Arrivals Handling

Late arrivals to any scorecard can present an infinite gamification design problem. Should they start with a zero score, and then have to catch up?

What if catching up is impossible? This can disincentivize participation. Not everyone starts scoring at the same time.

Here are four "handling late arrivals" design solutions:

1. **Cohort sharding**: bunch newcomers into cohorts in which all have an equal chance and time to succeed.
2. **Score resets**: everybody's score resets periodically, i.e., each week they start again at zero.
3. **Catch up ladders**: offer late arrivals a way to catch up with leader scores, perhaps with special bonuses.
4. **Divisions**: structure the score in different divisions so you only compare with people at the same stage as you.

Anti-Gaming Mechanics

With any long-term program there is time for gamers to figure out the system and "game it". This is typically where they find the cheapest way to win.

When you design, you need to design out unwanted gamer behaviours. Here are twelve anti-gaming mechanics you can deploy:

1. **Opaque scores**: hide the full details of the scoring system.
2. **Metric caps**: restrict capacity to score unlimited points.
3. **Penalty points**: dock points for bad behaviour.
4. **Reciprocity**: reward the response not the activity.
5. **Enforce a code of conduct**: set clear legal controls around the game.
6. **Remove financial rewards**: don't make winning prizes the sole aim.
7. **Peer transparency**: allow players to check each other's scores.

8. **Relative ranking method**: activity in every metric is required for a top score.
9. **Focus on intrinsic reward**: the main benefit is not the game.
10. **Stay the course**: continue the program after initial excitement has passed, the gamers may disperse naturally.
11. **Use score ratios**: ensure balanced behaviour across two or more metrics.
12. **Tweak metric weightings in mid-flight**: don't be afraid to change the rules until the mix of point allocation is correct.

    More established programs will need to signpost this in advance for fear of losing player trust. e.g., Google tells websites about impending search engine ranking algorithm changes well in advance.

"Don't blame the gamer,
blame the game"

Distribution Channel

Your choice of channel will depend on your audience. The channel you choose will affect how your program is received.

Here's a checklist of channels to consider:

- Email
- Website
- Mobile App
- Push Notifications
- Text Message
- Big Screen TV
- Connected TV (at home)
- Smart Speaker (Alexa, Google Home)
- Business Dashboard (e.g., Geckoboard, Cyfe)
- Twitter (@reply, direct messaging)
- Social Media Post (e.g., Facebook, Yammer, LinkedIn)
- Scores website (e.g., Ranker, Listly, Tribalist, Rise.global)
- Instant Message (WhatsApp, Slack)
- Print (e.g., a letter in the post)
- Ring Bound Report

- Giant Infographic
- Radio or TV News Broadcast
- Augmented Reality
- Virtual Reality

"The medium is the message"

–Marshall McLuhan

Program Framing

An infinite gamification program does not sit alone, it sits within a broader narrative that you communicate to your players.

Framing the score means explaining what it is, the benefits, and how you are expecting it to fit into player's lives. The best framing feels like an extension of what is already going on rather than a departure to somewhere new. So, use and extend existing language rather than try to apply a different metaphor, for example from a popular game. Think "Key Safety Points" rather than "Epic Dragon Hides".

The most important aspects of framing are:

- **Program name** - e.g., Sales Quality Program
- **Score name** - how you describe the total score - e.g., Rep Quality Score
- **League name**, (if you have one) - e.g., Reps Hitting Quality Benchmark This Week
- **Player name** e.g., Inside Sales Reps

Score Name

What do you call your score?

A checklist of useful words:

- Index
- Total
- Score
- Points

Try combining with words that summarise the behaviour you're looking for:

- Guru Index
- Power Score
- Experience Points
- Total Revenue

You can also borrow from sport (e.g., runs). But, be careful not to mix up scoring rules and terminology; you may confuse your players.

Player Name

What do you call your scorecard players? Chivvy them along with an encouraging label. They may yet live up to it!

- Users
- Stars
- Legends
- Hotshots
- Players
- Champions
- Big Cheeses
- VIPs
- Leaders
- Super Stars
- Gurus
- Top Dogs
- Pioneers
- Advocates

- Aces
- Bigwigs
- Ninjas
- Rock Stars
- Influencers
- Inspirers

League Name

What do you call your competition?

- Sales Contest
- Premier League
- President's Club
- 100% Club
- Top 100
- Power 100
- Championships
- Leaderboard
- Tournament
- Rankings
- Competition
- League
- Challenge
- Derby
- Death Pool

- Cup
- Trophy
- Classic
- Open
- Scoreboard

"A rose by any other name
would smell as sweet"
— William Shakespeare

Different Score Formats

Scores come in all shapes and sizes, with and without decimal places.

Score formats to consider are:

- Number
- Time
- Percentage
- Currency
- Grade
- Quartile
- Emoji (happy to sad)
- Distance (e.g., feet and inches)
- Red, Amber, Green
- Progress Bar

- Colour coding[6]

There is also the option of a binary score format which is either achieved or not achieved.

- Tick in the box
- Badge
- Certificate
- Membership

[6] While included here for completeness, I usually advise gamification designers not to use colour as a score indicator in their programs and certainly never as the only indicator. This is because colour meaning is culturally dependent and a proportion of your player population is likely to be colour blind.

Leaderboard Layout

When laying out your leaderboard for visual consumption, it's important to get the information hierarchy right first:

Let's Supercharge Our Sales ········ *Program name*

Week 17 - (12 April - 19 April) ········ *Score Period*

Team: North West ▼ ········ *Leaderboard name*

Rank	Player	Calls	InMails	Emails	SalesScore
1. ▲(3)	Jane Smith	12	4	18	**34**
2. ▼(1)	John Doe	3	3	20	**26**
3. ▼(2)	Jane Doe	16	2	7	**25**
4. ▶(4)	John Smith	9	1	5	**15**

Change in Rank · *Previous Rank* · *Underlying metrics* · *Total score highlighted in bold*

Be sure to resist the temptation to make the columns sortable. The essence of a leaderboard is that it lists players in the order you want people to focus on. Allowing sorting, for example of the Emails column, draws attention away from the status reward that your ranking brings players.

Scorecard Layout

In order to optimise performance, players need to be able to drill down into their score and see how they performed on underlying metrics. They want to compare results in each metric over time.

Let's Supercharge Our Sales

John Smith Scorecard

	Week 17 (12 April - 19 April)	*Week 16 (5 April - 12 April)*
Calls	9 ▼	*11*
InMails	1 ▲	*0*
Emails	5 ▲	*3*
SalesScore	**15** ▲	*14*
North West Rank:	4. ▶	*4.*
	Current Score Period	Previous Score Period

The key to a good scorecard is recognising that the score comes first - everything else, leader-board rank, team performance, badges earned and personal bests are derived from the score so should be displayed afterward.

Badge Design

Badges can be used in infinite gamification as long as they can be won multiple times, however this can devalue their scarcity value.

Examples might be "Player of the Match" or "Sales Rep of the Month".

Here are some elements you can include in your design:

Visual Elements	Name Elements
Border	Awarding Institution
Banner	Time Period
Icon	Award Date
Typography	Achievement
Colour	Level
Decoration	Category
Rating Stars	Qualifier (e.g., geo region, demographic - gender, age)

EVOLUTION

Infinite gamification programs iterate and change over time. The best programs have processes in place to manage change.

Without evolution, programs can ossify, become brittle and eventually break - resulting in unwanted side effects and disengagement.

Maturity Model

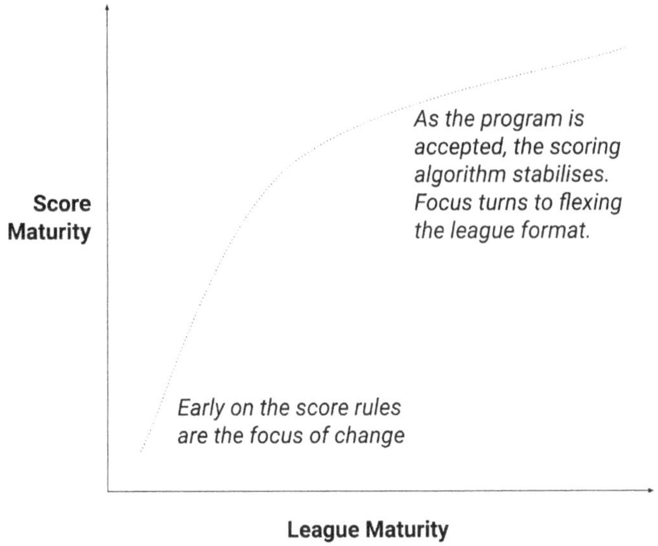

Score Maturity
As the program is accepted, the scoring algorithm stabilises. Focus turns to flexing the league format.
Early on the score rules are the focus of change
League Maturity

In the early stages of any scorecard, the "rules of the game"—how points are scored—require lots of change and tinkering.

In the later stages, the league structure—how individuals and teams are compared against others—changes, requiring updating and modification.

Finally, the scorecard and league settle down into the way "scores are kept 'around here'," and "the way we run our league".

Metric Maturity

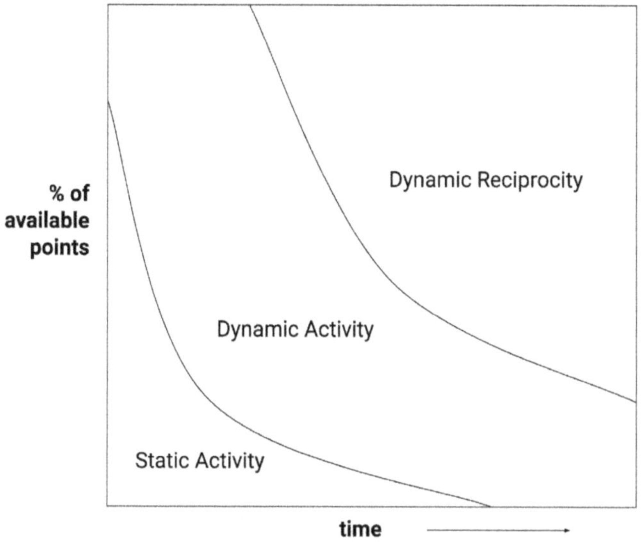

Over time the way metrics are counted in any score algorithm may change.

In the **early** stage, the bulk of points will be for Static Activity—one-time achievements, such as installing a new piece of software, onboarding or attending a training course.

In the **middle** stage, the bulk of points will be for Dynamic Activity—regular achievements that need to be done weekly or monthly (e.g., posting a blog article, or adding sales contacts).

In the **late** stage the bulk of points will be for Dynamic Reciprocity—regular achievements that others do as the result of activity (e.g., earned comments on a blog article, or customer purchases).

You can plan this ahead by creating a **road map** of metrics that will be introduced over time as the program develops.

"When a measure becomes a target, it ceases to be a good measure."

Goodhart's Law

Player Maturity

Players react to being given a score in different ways. How they react reflects their maturity in dealing with scorecards.

Baby	Accept the score as given
Toddler	Reject the score out of hand
Teenager	Attempt to game the scorecard
Adult	**Use the score to track progress and optimise activity**
Senior	Challenge the score algorithm itself
Retiree	Outgrow the score

It's important to recognise the difference between Toddler behaviour—someone who rejects the score without understanding it—as opposed to a Retiree, who rejects the score because it is no longer relevant to them. Toddler's may think they have the wisdom of Seniors, but as in real life, it is rarely true!

The Committee

Whatever you call it (scoring committee, management committee, cabinet, troika, dictatorship) iteration is often a balance of competing interests.

Meeting as a **scoring committee** allows you to regularly evaluate the effectiveness of the score algorithm, the league structure, and to recommend changes based on program feedback.

When you make changes be sure to stagger them and communicate them well in advance of the change.

"That's a matter for the committee!"

*"We **are** the committee."*

– Chariots of Fire

The points scoring system of Rugby Union has changed several times over the past 130 years. In 1886 you scored 1 point for a try and 2 for a conversion. Now it's 5 points for a try and 2 for a conversion—a complete reversal!

FINAL WORD

In this book we've covered:

- An **introduction** to infinite gamification
- How to **analyse** the landscape and demand for your program
- How to **design** a program that channels the right behaviours
- How to **evolve** your program over time so it continues to be relevant and useful

All that remains is for you to put it into practice. I'm super excited to hear what you come up with.

If you've got screenshots or testimonies of how this book has helped you and your organisation then please do share, I'd really love to hear your experience. Post online using the hashtag #infinitegamification. My screen handle is @tobyberesford on Twitter and pretty much everywhere else.

If you have specific questions related to your project and want expert advice, then please do ask them on Quora via the Gamification topic where a group of gamification gurus are usually around to answer.

If you're looking for a software solution to display your scorecards, leagues, and to notify your players then be sure to check out my scorebook platform Rise at www.rise.global.

Thank you – to Richard Beresford, Sophia Pope, Jonathan Peters, Monica Cornetti, Yu-kai Chou, Nicolas Babin, Anthony Beresford, Nick Shah, John Conaghan, Pete Jenkins, Alexandre Dutarte, Rob Alvarez, Bruno Ribeiro and Timlynn Babitsky who have all contributed to optimise this book and make it really quite a lot better than if I'd done this alone. Thank you!

Finally, special thanks to Amber, my wife, who I thank from the bottom of my heart for her belief in the vision, her patience and her willingness to support the many hours that went into both Rise and now this book. This book is dedicated to her and our three amazing children.

Further Reading

Everything assembled here owes a giant debt to the work of others whether in books, meetings, conferences, emails or social media posts.

The following books and authors have been valuable guides for my infinite gamification journey, so I list them here for you too:

- Coonradt, Charles A. *Game of Work: How to Enjoy Work as Much as Play*. Gibbs Smith Publisher, 2012.
- Coonradt, Charles A, Lee Benson, and Inc Game of Work. *Scorekeeping for Success*. Park City, Utah: Game of Work, Inc., 1998.
- Burke, Brian. *Gamify: How Gamification Motivates People to Do Extraordinary Things*. Brookline, MA: Bibliomotion, books + media, 2014.
- Pink, Daniel H. Drive: *The Surprising Truth about What Motivates Us*. Paperback ed. Edinburgh: Canongate, 2011.

- Zichermann, Gabe, and Christopher Cunningham. *Gamification by Design: Implementing Game Mechanics in Web and Mobile Apps.* Sebastopol, CA: O'Reilly, 2011.

Further Resources

Additional resources including videos and downloads can be found at:

www.infinitegamification.com

www.ingramcontent.com/pod-product-compliance
Lightning Source LLC
Chambersburg PA
CBHW070943210326
41520CB00021B/7029
* 9 7 8 1 8 3 8 0 4 5 3 3 3 *